"Hidden Wonders"

A Journey through the Curiosities and Discoveries of the Earth"

By

Brian Yout

Introduction

Welcome to an extraordinary journey, an exploration of the most remote frontiers of human knowledge. In these pages, we will seek to illuminate the dark recesses of the universe, uncover the intricate webs of our genetic code, peer into the beating heart of the atom, and probe the ethical challenges and promises of a future powered by nuclear energy.

"Beyond the Stars" is an invitation to dive into the mysteries of space exploration, a guide through the intricate maze of human genetic foundations, an in-depth examination of the potentials and complexities of nuclear energy, and a profound reflection on the quantum reality that permeates the fabric of our universe.

On this journey, we will delve into the unknown boundaries of our existence, exploring not only the vastness of space but also the minute details of our very essence. Guided by curiosity and the relentless pursuit of truth, we will delve into the depths of the unknown, seeking answers to age-old questions and raising new questions that will stimulate the mind and heart.

Get ready for an intellectual odyssey, where science intertwines with wonder, technology merges with philosophy, and humanity stands at the crossroads of extraordinary possibilities. "Beyond the Stars" is more than a book; it is an invitation to explore, to learn, to dream. We hope these pages inspire you as much as they inspired us in creating them.

TABLE OF CONTENT

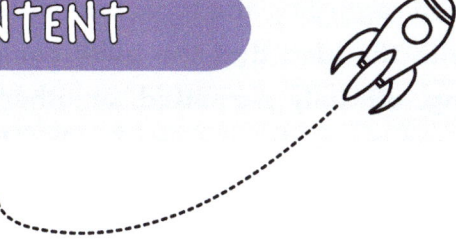

5.................................The Bermuda Triangle
6...Whale Song
8...Hessdalen Lights
10...............................The Great Barrier Reef
12..................................The Silence Triangle
14...................Costa Rican Stone Spheres
16..Namib Desert:
18..Nazca Lines:
20...................Hamza Underground River:
22..............Mystery of Disappearing Bees:
24......THE TEN WONDERS OF SCIENCE
25..DNA Structure:
27...................Einstein's Theory of Relativity:
29.................................Bohr's Atomic Model:
30.............................Discovery of antibiotics
32......................Darwin's Theory of Evolution
34............Standard model of elementary particles
36........................Quantum Mechanics Theory
38...Nuclear energy
40.............................Mapping the Human Genome
43...Space exploration
45....Conclusions: At the End of an Extraordinary Journey
46..Thanks

© Copyright 2024 by - Brian Yout - All rights reserved.

This document is geared towards providing exact and reliable information regarding the topic and issue covered. The publication is sold with the idea that the publisher is not required to render accounting, officially permitted, or otherwise qualified services. If advice is necessary, legal or professional, a practiced individual in the profession should be ordered.

From a Declaration of Principles which was accepted and approved equally by a Committee of the American Bar Association and a Committee of Publishers and Associations.

In no way is it legal to reproduce, duplicate, or transmit any part of this document in either electronic means or in printed format. Recording of this publication is strictly prohibited and any storage of this document is not allowed unless with written permission from the publisher. All rights reserved.

The information provided herein is stated to be truthful and consistent, in that any liability, in terms of inattention or otherwise, by any usage or abuse of any policies, processes, or Instructions contained within is the solitary and utter responsibility of the recipient reader. Under no circumstances will any legal responsibility or blame be held against the publisher for any reparation, damages, or monetary loss due to the information herein, either directly or indirectly. Respective authors own all copyrights not held by the publisher.

The information herein is offered for informational purposes solely and is universal as so. The presentation of the information is without contract or any type of guarantee assurance.

The trademarks that are used are without any consent and the publication of the trademark is without permission or backing by the trademark owner. All trademarks and brands within this book are for clarifying purposes only and are the owned by the owners themselves s, not affiliated with this document.

The Bermuda Triangle

The Bermuda Triangle is an area in the Atlantic Ocean forming a triangle between Miami (Florida, USA), Bermuda, and Puerto Rico. This region has become famous due to numerous cases of mysterious disappearances of ships, aircraft, and people over the years. These unexplained events have fueled various theories and legends, although most incidents can be reasonably explained through natural and human phenomena.

Some of the most popular theories include:

1. **Ocean currents and weather conditions:** The area is subject to strong ocean currents and unpredictable weather conditions, such as tropical storms. Some argue that these conditions may contribute to maritime and aviation incidents.
2. **Extraterrestrial activity:** Some believe that the Bermuda Triangle is a site of unidentified flying object (UFO) sightings and alien abductions. However, there is a lack of concrete evidence to support these claims.
3. **Anomalous magnetic fields:** Some suggest that the area may have anomalous magnetic fields that interfere with navigation and communication systems, leading to accidents.
4. **Atlantis and lost ancient civilizations:** Speculation exists that remnants of lost ancient civilizations, such as Atlantis, may influence the area with mystical energies or unknown technologies, causing disappearances and loss of contact.

Whale Song

Whale song is a complex and distinctive form of communication used by various whale species. This behavior has been primarily studied in humpback whales, but also in other species such as blue whales and gray whales.

Here are some key points about whale song:

Song Structure: Whale song is characterized by repetitive and melodic sequences of sounds. These sequences consist of a variety of sounds, such as gurgles, trills, and moans, organized into distinct patterns.

Duration and Complexity: Whale songs can last from a few minutes to several hours. The patterns can be incredibly complex and can vary significantly among whale populations. Additionally, songs can evolve over time, with whales adding new elements or modifying existing sequences.

Social Communication: It is believed that whale song serves a social communication function within populations. Studies suggest that males often sing to attract females during the mating season. The song might also be used to establish social hierarchies, communicate the location of food, or signal dangers.

Regional Specificity: Interestingly, whale songs vary considerably among different populations, and these variations can be so distinctive as to identify a specific population. In other words, whales from one region may have a characteristic song that sets them apart from other populations.

Evolutionary Adaptation: Whale song is considered an evolutionarily adaptive behavior that has contributed to the reproductive success of these creatures. The complexity of the song suggests involvement in sexual selection processes and in the formation and maintenance of social bonds within the whale community.

Despite scientists' efforts to understand the exact meaning of these songs, many facets of whale song remain a mystery, and research continues to explore the intricacies of this fascinating aspect of animal behavior.

Hessdalen Lights

Hessdalen Lights: In the Hessdalen valley in Norway, mysterious luminous phenomena occur in the form of spheres of light that manifest irregularly.

While some attribute these lights to natural phenomena like electrical discharges or gases, their precise origin remains uncertain, fueling speculation and research.

These luminous anomalies in Hessdalen have been observed since the early 1980s, and their unpredictable appearance has captivated the attention of scientists and enthusiasts alike. The lights often display vibrant colors and have been reported to move in ways inconsistent with known atmospheric phenomena.

Numerous scientific expeditions and studies have been conducted to unravel the mystery of the Hessdalen lights. Some researchers propose that ionized gases in the atmosphere, combined with particular geological features in the valley, may be responsible for the luminous displays. Others suggest that the lights could result from a unique combination of factors, including charged particles, electromagnetic interactions, and thermal variations.

Despite various hypotheses, a conclusive explanation for the origin of the lights has remained elusive. The intermittent nature of the phenomena adds to the challenge of studying them systematically.

The Hessdalen lights continue to be an intriguing enigma, prompting ongoing investigations and contributing to the allure of unexplained phenomena in the natural world. Scientists remain committed to solving the puzzle of these luminous manifestations in the Norwegian valley, balancing the quest for knowledge with the enduring allure of the unexplained.

The Great Barrier Reef

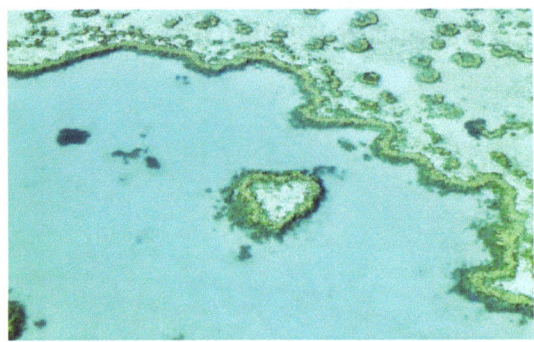

This vast coral system off the coast of Australia is one of the natural wonders of the world. Its incredible biodiversity and the complex ecological network still hold many undisclosed secrets, with scientists striving to better understand the dynamics governing this delicate marine ecosystem.

Stretching over 2,300 kilometers (1,430 miles), the Great Barrier Reef is a breathtaking mosaic of coral formations, teeming with an unparalleled diversity of marine life.

The intricate relationships between corals, fish, invertebrates, and other organisms create a finely tuned balance within this underwater world.

One of the enduring mysteries of the Great Barrier Reef lies in its resilience and adaptability. Despite facing threats such as coral bleaching, climate change, and human activities, the reef has displayed a remarkable ability to recover and withstand environmental challenges. Scientists are delving into the mechanisms that allow the coral organisms to adapt to changing conditions and exploring potential conservation strategies to preserve this natural wonder for future generations.

Additionally, the depths of the reef still harbor undiscovered species and unexplored regions, adding an air of mystery to this vibrant underwater ecosystem. Advanced technologies, including underwater drones and remote sensing devices, are being employed to unveil the hidden realms and better comprehend the intricate interactions that contribute to the reef's ecological richness.

While the Great Barrier Reef is celebrated for its beauty and ecological significance, the ongoing research and exploration underscore the urgency of preserving this global treasure. Scientists, environmentalists, and policymakers collaborate to implement sustainable practices and protective measures, ensuring the longevity and vitality of the Great Barrier Reef and unlocking the secrets held within its coral expanse.

The Silence Triangle

The Silence Triangle: A remote area in the Himalayas said to be characterized by an unusual lack of sounds. The absence of natural noises and the peculiar behavior of birds in this region make it a unique mystery, with explanations ranging from the particular topography to the unique acoustics of the area.

Nestled in the remote expanse of the Himalayas, the Silence Triangle is renowned for its uncanny absence of sounds. Within this isolated region, nature's usual cacophony is conspicuously muted, creating an atmospheric mystery that has captivated explorers and researchers alike. The distinctive hush is accompanied by the peculiar behavior of birds, adding layers to the enigma.

Scientists and theorists have proposed various explanations for this auditory anomaly. Some attribute the unusual quietude to the specific topography of the area, postulating that geographical features might influence sound propagation or absorption.

Others delve into the possibility of unique acoustics, where atmospheric conditions and natural formations conspire to create a rare auditory phenomenon.

As adventurers venture into the Silence Triangle, the tranquility enveloping the landscape sparks curiosity and raises questions about the intricate interplay between nature and sound.

This remote Himalayan mystery continues to beckon those eager to unravel the secrets hidden within its serene yet perplexing ambiance.

Costa Rican Stone Spheres

Costa Rican Stone Spheres: Scattered across various sites in Costa Rica, ancient stone spheres dating back thousands of years have puzzled archaeologists and enthusiasts alike. The enigma surrounding these perfectly round artifacts extends beyond their age, delving into the mysteries of their creation, transportation, and purpose.

Crafted with meticulous precision, the stone spheres range in size from a few centimeters to over two meters in diameter. Their near-perfect roundness and smooth surfaces have led to questions about the advanced craftsmanship and tools employed by the ancient civilization that created them.

The method of transportation poses another intriguing puzzle. Some of these massive spheres are located in seemingly inaccessible areas, raising questions about how a pre-Columbian society managed to move and position these colossal stones with such accuracy.

Speculation abounds regarding the purpose of these stone spheres. Some theories suggest they held ceremonial significance or served as markers for astronomical events. Others propose that the spheres played a role in societal rituals or had navigational purposes.

As researchers continue to investigate these archaeological marvels, the Costa Rican stone spheres remain shrouded in mystery, offering glimpses into the ingenuity of a civilization that left behind enigmatic artifacts, sparking fascination and inspiring ongoing exploration.

Namib Desert:

Namib Desert: Dominated by awe-inspiring and distinctively red sand dunes, the Namib Desert stands as a captivating testament to the wonders of nature.

While the striking landscape has drawn admiration for centuries, the processes behind the formation of these towering dunes and their remarkable crimson hue continue to be subjects of intensive study and scientific debate.

The Namib Desert's dunes, some of which are among the tallest in the world, are sculpted by the complex interplay of wind, sand, and geological factors. The prevailing theory suggests that these colossal dunes are shaped by consistent winds that carry grains of sand from the vast desert floor and deposit them into towering structures. The dunes' vibrant red color is attributed to the high iron oxide content in the sand, creating a stunning visual contrast against the desert's arid backdrop.

However, the exact mechanisms governing the formation of these dunes, their stability over time, and the precise role of environmental factors continue to be explored by scientists. Advanced technologies, including satellite imaging and computer simulations, contribute to unraveling the mysteries of the Namib Desert's ever-shifting sands.

As researchers delve into the intricacies of this mesmerizing desert landscape, the Namib continues to be a source of inspiration and scientific inquiry, offering a glimpse into the dynamic forces that shape our planet's most captivating terrains.

Nazca Lines

Nazca Lines: Etched into the arid expanse of the Nazca Desert in Peru, the Nazca Lines are colossal geoglyphs portraying intricate geometric figures and animals, their full grandeur only truly discernible from an aerial perspective.The mystery surrounding their creation, purpose, and the immense effort invested in their construction has intrigued scholars and adventurers alike for decades.

Believed to have been created by the Nazca people between 500 BCE and 500 CE, these massive drawings encompass intricate designs such as spiders, monkeys, birds, and geometric shapes, spanning vast areas of the desert floor. The precision and scale of these figures raise profound questions about the Nazca civilization's engineering prowess and their ability to comprehend such vast designs without an aerial viewpoint.

Numerous theories attempt to unravel the enigma of the Nazca Lines. Some scholars propose ceremonial purposes, suggesting that the Nazca people created these lines for rituals or religious ceremonies.

Others delve into the realm of astronomy, theorizing that the lines could have served as an elaborate astronomical calendar or markers for celestial events.

The exact methods employed to construct these colossal drawings, some of which stretch for hundreds of meters, remain a subject of ongoing investigation. The Nazca Lines continue to be a testament to the ancient ingenuity of the Nazca people, inviting speculation and admiration for their monumental efforts in creating these mysterious and visually stunning geoglyphs.

Hamza Underground River

Hamza Underground River: Unveiled beneath the vast expanses of the Atlantic Ocean, the Hamza Underground River stands as a colossal subaqueous discovery, challenging our understanding of the Earth's hidden dynamics. This immense freshwater mass, still shrouded in mystery, has sparked scientific intrigue and debate since its revelation.

Discovered in the early 2010s, the origin of the Hamza Underground River remains a subject of ongoing investigation.
Some scientists propose that it may originate from the melting of glaciers or subterranean aquifers, while others explore the possibility of a complex interplay between geological and hydrological processes.
The potential ramifications of the Hamza Underground River extend beyond its mere existence. Scientists hypothesize that this vast reservoir of freshwater could play a pivotal role in ocean circulation dynamics, influencing the larger climate system. Understanding the interactions between this hidden river and the broader marine environment is a critical endeavor, with implications for our comprehension of global climate patterns and their potential shifts.

As researchers delve into the depths of this subaquatic enigma, the Hamza Underground River stands as a testament to the ongoing discoveries that illuminate the intricate workings of our planet, emphasizing the interconnectedness of seemingly disparate elements in the Earth's complex and dynamic system.

Mystery of Disappearing Bees

Mystery of Disappearing Bees: The disconcerting phenomenon known as "colony collapse disorder" has cast a shadow over bee populations worldwide, posing a significant threat to the delicate balance of our food chain.

While the importance of bees in pollination is indisputable, the precise causes behind the enigmatic and widespread decline in bee populations remain elusive.

Colony collapse disorder is characterized by the abrupt and widespread disappearance of worker bees from their hives, leaving behind the queen and immature bees. This perplexing occurrence has been linked to a complex interplay of multiple factors. Pesticides, including neonicotinoids, are suspected to play a role, as these chemicals may adversely affect bee behavior, navigation, and immune systems. Diseases, such as various pathogens and parasites, contribute to the vulnerability of bee colonies.

Environmental changes, habitat loss, and climate fluctuations further compound the challenges faced by these vital pollinators. The intricate web of interactions among these factors makes it challenging to pinpoint a singular cause for the decline.

Scientists and researchers worldwide are diligently working to unravel the dynamics of this mysterious decline in bee populations. Collaborative efforts involve studying the impact of pesticides on bee health, developing disease-resistant bee strains, and implementing conservation measures to safeguard natural habitats for pollinators.

The urgency of understanding and addressing the mystery of disappearing bees extends beyond the apian realm. The decline of bee populations has far-reaching consequences for agriculture, ecosystems, and human food security. As ongoing research seeks to untangle the complexities of colony collapse disorder, the plight of bees remains a poignant reminder of the intricate balance within our interconnected natural world.

THE TEN WONDERS OF SCIENCE

DNA Structure

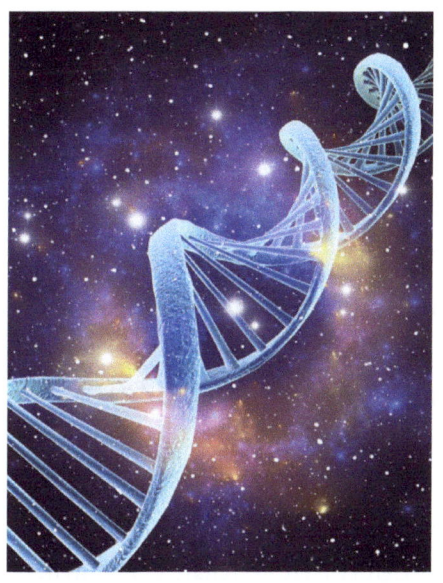

DNA Structure: The discovery of the double helical structure of DNA by James Watson and Francis Crick unveiled the way genetic information is organized and transmitted within cells, paving the way for understanding genetics and hereditary diseases.

- **Double helix:** The fundamental structure of DNA is in the shape of a double helix. This double helix consists of two long, thin strands wrapped around each other, similar to a spiral staircase.
- **Nucleotides:** The two strands of the double helix are made up of smaller molecules called nucleotides. Each nucleotide consists of three main components: a phosphate group, a sugar (deoxyribose in the case of DNA) and a nitrogenous base.
- **Nitrogenous bases:** There are four nitrogenous bases present in DNA nucleotides: adenine (A), thymine (T), cytosine (C) and guanine (G). These bases form specific pairs through hydrogen bonds: adenine always bonds to thymine, while cytosine always bonds to guanine. This complementarity of bases is critical for the stability of the double-helix structure.

- **Hydrogen bonding:** The nitrogenous base pairs bind to each other through hydrogen bonds, providing the stability necessary for the DNA double helix structure.
- **Antiparallelism:** The two strands of DNA run in opposite directions, meaning that while one strand proceeds from 5' to 3', the other proceeds from 3' to 5'. This antiparallel arrangement is crucial to the process of DNA replication.

- **Replication:** The structure of DNA allows for its replication during the process of cell division. During replication, enzymes break the hydrogen bonds between nitrogen base pairs, separate the strands and form new complementary strands using each strand as a template.

- In summary, the structure of DNA is a double helix formed by nucleotides, with each nucleotide containing a nitrogenous base. The complementarity of the bases and hydrogen bonds give stability to the structure, while the antiparallel arrangement and replication capability enable the transmission of genetic information from one cell generation to the next.

Einstein's Theory of Relativity:

Einstein's Theory of Relativity: Einstein's theory of relativity revolutionized physics by showing how space and time are linked to the presence of matter and energy. The theory predicted phenomena such as the curvature of space-time, later confirmed by astronomical observations.

First of all, Einstein suggested that space and time were not separate entities but closely linked, forming an interconnected space-time. Moreover, the presence of matter and energy not only influenced space and time, but determined their structure and geometry.

One of the most amazing aspects of the Theory of Relativity is the prediction of the curvature of space-time. According to Einstein, massive objects, such as planets or stars, "curve" space-time around them. This phenomenon is analogous to putting a heavy ball on an elastic surface, deforming it.

The theory also predicted the concept of time dilation and length contraction when moving at speeds close to that of light. These effects, although difficult to perceive at everyday speeds, have been confirmed by scientific experiments and have an important impact in technological applications, such as GPS satellites.

Subsequent astronomical observations have confirmed many predictions of the Theory of Relativity. For example, during a solar eclipse, the curvature of space-time predicted by the theory affected the light from stars close to the Sun, confirming the accuracy of Einstein's predictions.

Einstein's Theory of Relativity transformed the way we perceive space, time and gravity, showing that these quantities are not fixed and unchanging entities, but dynamic and interconnected. The theory continues to be fundamental in understanding cosmic phenomena and in guiding modern space exploration.

Bohr's Atomic Model

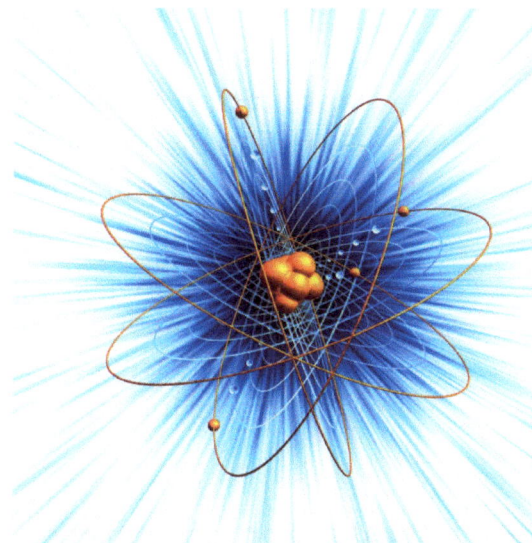

Bohr's Atomic Model: Bohr's atomic model provided a clear view of the structure of atoms, explaining how electrons orbit around the nucleus in specific energy levels. This understanding is essential to particle physics and chemistry.

Bohr atomic model: The Bohr atomic model, proposed by Niels Bohr, is a clear representation of the structure of atoms. According to this model, electrons orbit the nucleus in specific energy levels, or quantized orbits. Bohr introduced the idea that electrons occupy discrete orbits and that the energy of the electrons is quantized. This model contributed significantly to the understanding of atomic physics, explaining phenomena such as atomic spectra and providing an important conceptual basis for quantum theory. Bohr's vision has had a lasting impact on chemistry and particle physics, influencing our understanding of atomic structure and the fundamental principles governing the behavior of atoms.

Discovery of antibiotics

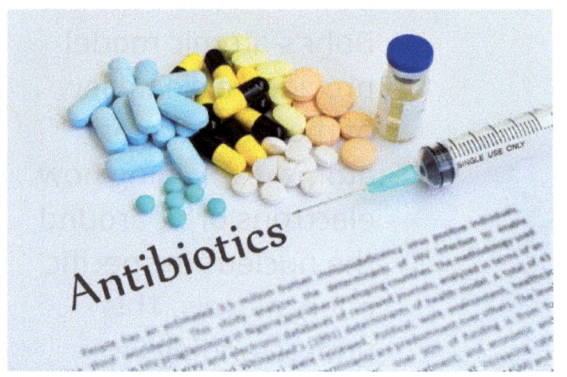

Discovery of antibiotics: Alexander Fleming's discovery of penicillin introduced antibiotics, revolutionizing medicine and enabling the effective treatment of bacterial infections.

Penicillin and Alexander Fleming: The discovery of antibiotics began with Sir Alexander Fleming in 1928, when he accidentally noticed that a mold called Penicillium produced a substance that killed bacteria. This substance was later called penicillin.

Effects of the antibiotic: Penicillin proved to be an effective weapon against a wide range of pathogenic bacteria. It was the first antibiotic to be used on a large scale to fight bacterial infections.

Antibiotic Era: The discovery of penicillin marked the beginning of the "antibiotic era," a revolution in medicine that made it possible to effectively treat many infectious diseases that in the past often led to death.

Large-scale production: Penicillin was the first antibacterial substance to be produced on a large scale through industrial processes. This made it possible to treat a large number of patients and helped to significantly reduce the mortality rate due to infections.

Development of new antibiotics: After penicillin, many other antibiotics were discovered and developed. Among the best known are streptomycin, tetracycline, and later more modern antibiotics such as ciprofloxacin and amoxicillin.

Antibiotic resistance: Over time, the overuse and misuse of antibiotics has led to the development of resistant bacteria, creating a global challenge known as antibiotic resistance. This threat calls into question the effectiveness of antibiotics in treating infections and underscores the importance of responsible use of these drugs.

Medical applications: Antibiotics are used to treat a wide range of infections, including pneumonia, urinary tract infections, skin infections, and infected wounds. They are also critical in surgery to prevent and treat postoperative infections.

the discovery of antibiotics was a milestone in the history of medicine, leading to significant improvements in medical care and survival in the presence of bacterial infections. However, responsible management of antibiotics is crucial to preserving their effectiveness over time.

Darwin's Theory of Evolution

Darwin's Theory of Evolution: Charles Darwin's theory of evolution explained how species evolve over time through the process of natural selection, offering a clear view of biological diversity.

Charles Darwin: The theory of evolution was formulated primarily by Charles Darwin, an English naturalist, in his 1859 book titled "The Origin of Species."

Natural selection: The key concept of the theory is natural selection, which holds that species evolve over time through the transmission of inherited characteristics that confer an adaptive advantage to the environment. Characteristics that are useful for survival are more likely to be passed on to subsequent generations.

Adaptation: The theory suggests that populations of organisms gradually adapt to their environment through natural selection. Individuals with favorable characteristics are more likely to survive and reproduce, passing those characteristics on to future generations.

Genetic variability: Genetic variability is fundamental to the theory of evolution. Genetic mutations and genetic recombination contribute to diversity within a population, providing the material for natural selection.

Common origin: Darwin proposed that all life forms on Earth share a common ancestor. This concept is known as the idea of the common origin of life and unified the diversity of life into an evolutionary family tree.

Fossils and paleontological evidence: Fossil evidence provides a record of organisms that lived in past ages, documenting evolutionary changes over time. Fossils have helped establish connections between current and ancient life forms.

Comparative anatomy: The theory of evolution is supported by comparative anatomy, which examines similarities and differences in anatomical structures between different species. The presence of homologous organs (similar structures having a common ancestor) suggests an evolutionary link.

Biogeography: Analysis of the geographical distribution of species has provided further evidence in favor of the theory of evolution. For example, Darwin observed that similar species were often found in nearby geographical locations.

Molecular evidence: With the development of molecular biology, evidence at the molecular level, such as DNA and protein sequences, confirmed evolutionary relationships among species and supported the idea of common origin. Darwin's theory of evolution is one of the pillars of modern biology and has had a profound impact on our understanding of the diversity of life on Earth.

Standard model of elementary particles

Standard model of elementary particles: The standard model describes the fundamental particles that make up matter and the forces acting on them, contributing to the understanding of subatomic interactions.

Description of fundamental particles: The Standard Model is a theory that describes the fundamental particles that make up matter. These particles are mainly divided into q**uarks (constituents of protons and neutrons) and leptons (such as electrons and neutrinos).**

Fundamental forces: The Standard Model identifies four fundamental forces that govern the interactions between particles. These forces are the electromagnetic force, the strong nuclear force, the weak nuclear force and the gravitational force.

Gauge bosons: To describe the fundamental forces, the Standard Model introduces particles called gauge bosons. For example, the photon is associated with the electromagnetic force, the gluon with the strong nuclear force, and the W and Z bosons with the weak nuclear force.

Higgs boson: A key component of the Standard Model is the Higgs boson, discovered in 2012 at the Large Hadron Collider (LHC). The Higgs boson explains how particles acquire mass, providing a complete picture of the origin of mass in our universe.

CKM Matrix: The Standard Model incorporates the CKM (Cabibbo-Kobayashi-Maskawa) Matrix, which describes the transitions between different types of quarks. This matrix is important for understanding the decay processes of quarks.

Quantum Chromodynamics (QCD): The Standard Model includes QCD, a theory that describes the interaction of quarks through the transport of particles called gluons. QCD explains the strong nuclear force responsible for the cohesion of protons and neutrons in atomic nuclei.

Quantum electrodynamics (QED): QED is an essential part of the Standard Model and explains the electromagnetic interaction between charged particles, such as electrons and positrons.

Unification of forces: One of the aspirations of the Standard Model is the possible unification of fundamental forces. Although this has not yet been fully achieved, the model has brought descriptions of electromagnetic and weak nuclear forces closer together.

Experimental verifications: The Standard Model has withstood numerous experimental verifications. However, some issues, such as the nature of dark matter and the non-inclusion of the force of gravity, indicate that it may not be the final theory of physics.

The Standard Model represents the most accurate description of fundamental particles and the forces acting on them, contributing significantly to our understanding of subatomic interactions.

Quantum Mechanics Theory

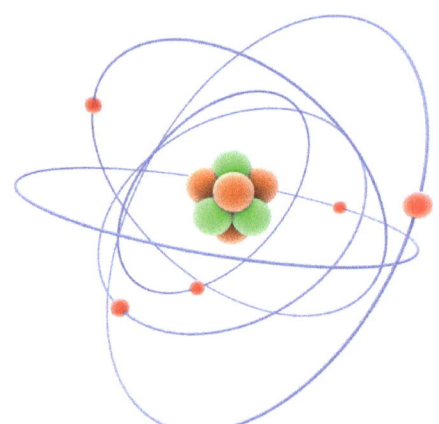

Quantum mechanics theory: Quantum mechanics has transformed our understanding of matter at the subatomic level, introducing principles such as indeterminacy and wave-particle duality.

Birth of Quantum Mechanics: Quantum Mechanics emerged in the early 20th century as a result of experiments with phenomena at atomic and subatomic scales that could not be explained by classical physics.

Principle of wave-particle duality: One of the key concepts of Quantum Mechanics is wave-particle duality. Subatomic particles, such as electrons, can exhibit both particle and wave behaviors, depending on experimental circumstances.

Heisenberg indeterminacy: The Heisenberg indeterminacy principle states that we cannot simultaneously know precisely the position and momentum of a subatomic particle. This principle has profoundly influenced our understanding of the quantum world.

Quantum states and wave function: Quantum Mechanics describes the state of a particle through a wave function, a mathematical representation that contains information about the probability of finding the particle at a certain position or with a certain momentum.

Quantum superposition: According to Quantum Mechanics, particles can exist in superposition states, which means that they can simultaneously occupy multiple possible states as long as they are not observed.

QQuantization of energy: In Quantum Mechanics, energy is quantized, which means that it can take on only discrete values. This explains phenomena such as emission and absorption spectra in atomic systems.

Principle of superposition and decoherence: The principle of superposition states that a particle can exist in several states at once, but when observed, the system collapses into a particular state. Decoherence is the process by which this transition occurs in practice.

Technological applications: Quantum Mechanics has led to fundamental technological developments, such as the creation of transistors, lasers, and advanced imaging techniques such as nuclear magnetic resonance (NMR) and positron emission tomography (PET).

Interpretations of Quantum Mechanics: There are several interpretations of Quantum Mechanics, including the Copenhagen interpretation, the quantum decoherence interpretation, and the Many-Worlds interpretation. Each proposes a different view of quantum reality.

Quantum Mechanics has radically transformed the way we understand the subatomic world, challenging the insights of classical physics and paving the way for new technologies and scientific discoveries.

Nuclear energy

Nuclear energy: Nuclear energy uses nuclear reactions to produce energy, influencing both electricity production and medicine with radiation therapy.

Energy Source: Nuclear energy is a source of energy obtained by processes involving the nucleus of atoms. The two main applications are nuclear energy for electricity generation and medical and industrial applications.

Nuclear fission: The primary source of nuclear energy is nuclear fission, a process in which the nucleus of an atom splits into two or more smaller nuclei, releasing energy. This process is the basis of nuclear power plants.

Nuclear reactors: Nuclear reactors are facilities designed to harness nuclear fission in a controlled manner. Inside a reactor, nuclear fuel (usually uranium or plutonium) undergoes fission, generating heat that is converted to electricity.

Energy advantages: Nuclear power offers a highly concentrated source of energy, with a large amount of energy produced from a small amount of nuclear material. Nuclear power generation is also continuous and can generate large amounts of electricity efficiently.

Greenhouse gas emissions: Nuclear power plants produce electricity without emitting greenhouse gases during the power generation process, unlike some conventional fossil fuel-based power sources.

Nuclear waste: One of the main challenges of nuclear power is the management of radioactive waste produced during the nuclear fission process. This waste requires safe storage for long periods of time.

Nuclear accidents: Incidents such as those at Three Mile Island, Chernobyl and Fukushima have highlighted the risks associated with nuclear power. Concerns include the spread of radiation and potential health and environmental effects.

Nuclear energy in the medical sector: In addition to electricity generation, nuclear energy is used in medicine for diagnosis and treatment, such as through positron emission tomography (PET) and radiation therapy.

Development of advanced reactors: Ongoing research to develop advanced nuclear reactors, such as nuclear fusion reactors, aims to overcome some of the challenges and risks associated with traditional nuclear power.

Nuclear energy debate: Nuclear energy is being debated on ethical, environmental and safety issues. Some emphasize its role in providing low-carbon energy, while others worry about the associated risks and stress the importance of renewable alternatives.

Nuclear power is a source of energy that has advantages and challenges, and its role in electricity generation and industrial applications is being discussed in depth around the world.

Mapping the Human Genome

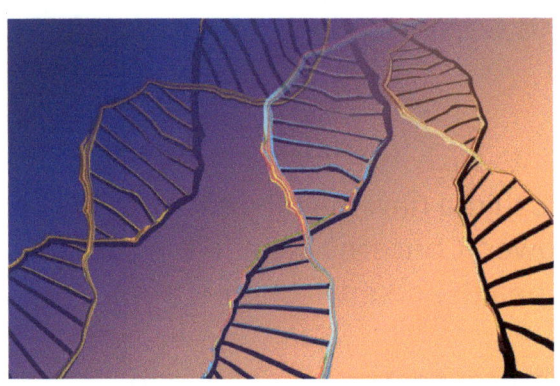

Mapping the Human Genome: Mapping the human genome has made it possible to identify all the genes in human DNA, paving the way for new breakthroughs in personalized medicine and understanding genetic diseases.

Human Genome Project (HGP): Mapping the human genome was an ambitious international scientific project initiated in the 1990s, known as the Human Genome Project. The goal was to sequence and map the entire human genome, that is, the complete set of genes found in human cells.

DNA sequencing: Mapping the human genome involved DNA sequencing, identifying the order of nucleotides in DNA base pairs. This involved determining the sequences of more than 3 billion base pairs in human DNA.

Complete Mapping of the Human Genome: In 2003, the Human Genome Project successfully announced the complete mapping of the human genome, making available a high-quality reference sequence.

Human genes and genetic functions: Genome mapping has made it possible to identify and catalog approximately 20,000-25,000 human genes. This has paved the way for understanding the genetic functions associated with specific genes, including those related to inherited diseases.

Personalized medicine: Genome mapping has made possible the advent of personalized medicine. Knowing an individual's genetic profile can help predict responses to specific treatments and tailor therapies to each patient's genetic makeup.

Scientific research and biotechnology: Genome mapping has fueled a wide range of scientific research, helping to better understand the genetic basis of complex diseases and paving the way for new developments in biotechnology and gene therapies.

Human Genome and Evolution: Studying the human genome has also provided crucial information about our evolution as a species. It has made it possible to trace the genetic similarities and differences between humans and other organisms.

Ethical and legal challenges: Genome mapping has raised important ethical and legal issues, such as genetic privacy and access to genetic information. The scientific community and institutions faced regulatory and ethical challenges related to the use of genetic information.

Subsequent initiatives: Since the completion of the Human Genome Project, subsequent initiatives have emerged to deepen understanding of the human genome, such as the ENCODE (Encyclopedia of Human Functional Elements) Project, which aims to identify and Understand the function of all functional elements of the human genome.

Impact on biomedical research: The mapping of the human genome has had a revolutionary impact on biomedical research, accelerating the discovery of disease-associated genes and opening up new perspectives for the development of targeted therapies

Space exploration

Space exploration: Space exploration missions have provided detailed images and information about planets, moons and galaxies, contributing to our knowledge of the universe and the search for extraterrestrial life.

Human space exploration: Human space exploration involves sending astronauts into space to conduct various missions, including research, experiments and studying the effects of microgravity on the human body. The International Space Station (ISS) has been the focus of these activities.

Unmanned space missions: Unmanned space missions use robotic spacecraft to explore space, planets, moons and other celestial bodies. These missions provide valuable data and images without the need to send humans into space.

Moon landings: The Apollo program, initiated by NASA, succeeded in landing humans on the Moon for the first time in 1969 with the Apollo 11 mission. This historic achievement was followed by numerous other Apollo missions, which contributed to the knowledge of the Moon. This historic achievement was followed by numerous other Apollo missions, which contributed to our understanding of the Moon.

Mars exploration: Mars has been a focal point for exploration. Robotic missions such as rovers (Spirit, Opportunity, Curiosity, Perseverance) have been sent to study the Martian surface, geology, and search for signs of past or present life.

Interplanetary probes: Probes such as Voyager, Pioneer and New Horizons have been launched to study and collect data on planets, moons and other celestial bodies in our solar system and beyond.

International collaboration: Space exploration often involves collaboration between various countries and space agencies. The ISS, for example, is a multinational effort involving NASA, Roscosmos, ESA, JAXA, and CSA.

Private Space Companies: The 21st century has seen the rise of private companies, such as SpaceX, Blue Origin, and others, actively participating in space exploration. They have developed reusable rocket technologies and aim to make space travel more cost-effective.

Hubble Space Telescope: Launched in 1990, the Hubble Space Telescope has provided stunning images and valuable data, contributing to our understanding of distant galaxies, nebulae, and the age of the universe.

Search for Extraterrestrial Life: Scientists use space telescopes and other instruments to search for signs of extraterrestrial life by studying exoplanets, moons, and environments that may harbor the conditions for life.

Future Missions: Upcoming missions include plans to return humans to the Moon (Artemis program), send humans to Mars, explore asteroids, and launch space telescopes like the James Webb Space Telescope.

Space Tourism: Emerging trends in space exploration include the potential for space tourism, with private companies working on commercial spaceflight experiences for civilians.

Conclusions: At the End of an Extraordinary Journey

In this extraordinary journey through the pages of "Hidden Wonders", we crossed continents and oceans, exploring the depths of the Earth and peering into the infinite sky. We experienced first-hand the diversity of wildlife, ventured to iconic and hidden places, and traveled through time, embracing the traces of Earth's history.

In this conclusion, we reflect on the wealth of discoveries and wonders we have shared. Each page brought us closer to understanding the intrinsic beauty of the Earth and the incredible complexity of its mechanisms. We have learned that behind every curiosity there is a story, and in every discovery there is a world to explore.

This book was conceived as an ode to human curiosity, an invitation to look beyond the surface and embrace the wonder that lies in every corner of our planet. The stories told here are just a taste of the countless adventures that await those who are ready to dive into the unknown with open eyes and hearts eager to learn.

As we conclude this journey, let's keep the flame of curiosity alive. May every unresolved question be an invitation to new explorations, and may every wonder encountered fuel our thirst for knowledge. The Earth, with its hidden wonders, is an extraordinary gift that offers us infinite secrets to discover and holds treasures that patiently await us.

Thank you for sharing this adventure with us. May your spirit of discovery continue to guide you, and that the wonders of the Earth will always accompany you on your personal journey through knowledge and beauty.

Thanks......................

I would like to express my sincere gratitude to all the people who contributed and supported during this project.

First, I would like to thank my family for their constant support and encouragement. Without your love and understanding, this journey would have been much more challenging.

Special thanks go to my friends and colleagues who shared valuable ideas, offered practical help and made this journey more enjoyable.

Warm thanks to all the people involved in my training and knowledge acquisition. Every teacher, professor, or mentor helped shape my educational journey.

Heartfelt thanks to the scientific and academic community, a source of constant inspiration and a place of fascinating discoveries.

Finally, I thank everyone who contributed directly or indirectly to this journey. Your support has been crucial, and I deeply appreciate every single contribution.

Thank you all for making this project an amazing and unforgettable experience.

Printed by Libri Plureos GmbH in Hamburg, Germany